PHYSICS FOR KIDS

49 Easy Experiments with Heat

Robert W. Wood,
Illustrated by Steve Hoeft

TAB Books

Division of McGraw-Hill, Inc.

New York San Francisco Washington, D.C. Auckland Bogotá
Caracas Lisbon London Madrid Mexico City Milan
Montreal New Delhi San Juan Singapore
Sydney Tokyo Toronto

pbk 3 4 5 6 7 8 9 10 11 DOC/DOC 9 9 8 7 6 5
hc   1 2 3 4 5 6 7 8   9 DOC/DOC 9 9 8 7 6 5 4 3 2 1 0

Library of Congress Cataloging-in-Publication Data

Wood, Robert W., 1933-
    Physics for kids : 49 easy experiments with heat / by Robert W. Wood ; illustrated by Steve Hoeft.
        p.   cm.
    Summary: Presents a collection of experiments exploring the properties of heat.
    ISBN 0-8306-9292-4   ISBN 0-8306-3292-1 (pbk.)
    1. Heat—Experiments—Juvenile literature.   [1. Heat-
-Experiments.   2. Experiments.]   I. Hoeft, Steve, ill.   II. Title.
QC256.W66   1990
536'.078—dc20   89-28588
                                              CIP
                                              AC

Acquisitions Editor: Kimberly Tabor
Book Editor: Lori Flaherty
Production: Katherine G. Brown
Paperbound cover photograph by Susan Riley, Harrisonburg, Va.

Contents

Introduction

PHYSICS IS A TERM THAT COMES FROM A GREEK WORD THAT MEANS NA-ture. It is the science that studies the "how" and "why" of the natural world around us. Physics explains why water freezes or evaporates, how a magnifying glass works, and why a tennis ball bounces.

The subjects in this exciting field often overlap, but they can be divided into several basic groups: mechanics, heat, light, electricity and magnetism, and sound.

Until about 1800, it was generally thought that heat was a fluid called caloric. Because an object weighed the same whether it was hot or cold, heat was a fluid of no particular importance. Then the idea that heat might be used to do work was considered. It was soon discovered that heat is a form of energy.

Heat affects most everything we do, what we wear and the types of houses we live in. If the sun should cool, all life on Earth would disappear and it would be just a cold, dead mass much like the moon. Heat is both our servant and our enemy. Using heat we can move trains, ships, and airplanes, but heat can also burn.

All materials or substances are made up of tiny particles called mol-ecules. It is thought that these molecules are in constant motion or vibration. Any moving body has kinetic energy. The faster it moves the more kinetic energy it has. The same happens with molecules. Because they are moving, they have kinetic energy called heat. The faster the molecules move the more kinetic energy and the more heat they have.

Because heat is a form of energy, any form of energy (electrical, chemical, or mechanical) can be changed into heat energy. Heat energy can also be changed into each of the other forms of energy.

Heat can travel from one place to another in three ways: conduction, convection, and radiation. Conduction is when heat moves from one mol-ecule to another molecule. This is where the molecules near the source of heat become hot, move faster, and strike the molecules next to them. These molecules then strike the molecules next to them, causing them to move faster and get hotter. This continues until the heat gradually moves along.

Convection is when heat is transmitted by currents of gases or liq-uids. Warm air from a stove is pushed up by the colder, heavier air around it. The air begins to circulate and warm the room.

Dark and light surfaces of the earth are heated to different temperatures by the sun. These warm and cool areas produce convection currents we call wind. Water in a pan circulates much the same way air does when it is heated. As the water is warmed, it becomes less dense than the cooler water surrounding it and rises.

Radiation is when heat travels in waves in all directions. In conduction and convection, heat must travel by moving particles, but heat can also travel where matter does not exist. This can be seen by the sun's heat warming the earth through several million miles of space. Heat waves are similar to light waves. Both are electromagnetic waves except heat waves are longer than light waves. Heat waves are sometimes called infrared rays.

All things have some heat. Water might be only a little warmer than an ice cube. Cold only means some of the heat has been removed. This book is an introduction into the fascinating area of physics called thermodynamics or heat. It is the study of how heat can do work, how it moves from one place to another, and how it affects other properties.

Conduct all of the experiments in this book with safety in mind. Be careful; some experiments require an adult to help you. If you are just the least bit unsure about anything, ask a parent or teacher for help or advice before you begin. Be sure to read the section on symbols used in this book that follows. These symbols will alert you to any safety precautions you should take and whether an adult should help you.

Symbols Used in This Book

MANY OF THE EXPERIMENTS USED IN THIS BOOK REQUIRE THE USE OF BURN-ing candles. It is recommended that a parent or teacher supervise young children and instruct them on the hazards of fire and how to extinguish flames safely. It is also recommended that children be advised on what to do in case of fire.

All of the experiments in this book can be done safely, but young children should be instructed to respect fire and the hazards associated with careless use. The following symbols are used throughout the book for you to use as a guide to what children might be able to do independently, and what they *should not do* without adult supervision. Keep in mind that some children might not be mature enough to do any of the experiments without adult help, and that these symbols should be used as a guide only and do not replace good judgment of parents or teachers.

 Flame is used in this project and adult supervision is required. Do not wear loose clothing. Tie hair back. When handling candles, wear protective gloves—hot wax can burn. Never leave a flame unattended. Extinguish flame properly. Protect surfaces beneath burning candles.

 The use of the stove, boiling water, or other hot materials are used in this project and adult supervision is required. Keep other small children away from boiling water and burners.

 Protective gloves that are flame retardant and heat resistant should be worn. Handling hot objects and hot wax can burn hands. Protect surfaces beneath hot materials—do not set pots of boiling water or very hot objects directly on table tops or counters. Use towels or heat pads.

 Materials or tools used in this experiment could be dangerous in young hands. Adult supervision is recommended. Be sure to keep tools and materials out of reach of young children after use.

Scissors are used in this project. Young children should be supervised carefully and older children instructed to exercise caution.

 Electricity is used in this experiment. Young children should be supervised and older children cautioned about the hazards of electricity.

 Protective safety goggles should be worn to protect against shattering glass, flying debris, and other hazards that could damage your eyes.

PART I

49 EASY EXPERIMENTS IN HEAT

Experiment 1

Materials

☐ 3 large bowls
☐ water (very warm, room temperature, and very cold)

Measuring Temperature with Our Hands

Align the bowls in a row and fill the first one with very cold water, the middle one with water at room temperature, and the one on the right with very warm water; almost hot, but not hot enough to burn you.

Place your left hand in the cold water and your right hand in the very warm water. Leave them in the water a few seconds. Now take your right hand from the very warm water and quickly place it in the middle bowl with the water at room temperature. Notice how it feels. It will feel much cooler than the temperature of the water actually is.

Next, remove your right hand and place your left hand from the cold

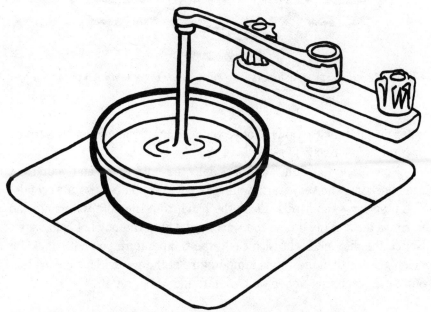

Fig. 1-1. *Fill the bowls about three-fourths full of water.*

Fig. 1-2. *Submerge your hands long enough for them to adjust to the temperature.*

water into the middle bowl. Notice the feeling; it will feel much warmer than the water really is.

Now place your left hand back into the cold water and your right hand back into the very warm water. Leave them there several seconds. Then quickly swap hands. Place the hand from the cold water into the warm water and the one in the warm water into the bowl of cold water. Notice the difference. In each case, the temperature you felt should be much greater than the actual temperature of the water. This means that our hands cannot be relied on to determine temperature.

Fig. 1-3. *The temperature changes seem much greater than they really are.*

Experiment 2

Materials

- [ ] thermometer
- [ ] jar of ice
- [ ] pot of boiling water

Testing a Thermometer

The scale on a thermometer is marked at two important points. The point where water turns to ice and the point where water turns to steam.

Place the thermometer down into the jar of ice. Leave it there a few

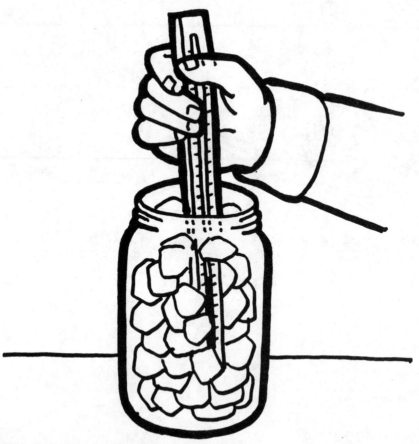

Fig. 2-1. *The freezing point of water is 32 degrees Fahrenheit.*

minutes. Notice how close it will read to 32 degrees Fahrenheit (F) or 0 degrees Centigrade (C).

Next, remove the thermometer and allow it to warm up some. Wearing protective gloves, place the end of the thermometer just above the surface of the boiling water. Be careful not to get burned. It might need to stay there a minute or two. It should read near 212 degrees F or 100 degrees C. Remove the thermometer and allow the end to cool before touching it.

Fig. 2-2. *Water boils and turns to vapor at 212 degrees Fahrenheit.*

Experiment 3

Measuring the Temperature of Air

Materials

- ☐ 2 thermometers
- ☐ 2 pieces of folded cardboard
- ☐ 2 rubber bands
- ☐ large jar
- ☐ sunny day

Loop the rubber band around each piece of cardboard and make two stands. Place a thermometer inside each stand. Place one stand into the jar then place both thermometers in bright sunlight. A window sill works fine. One thermometer should be in the jar and the other standing

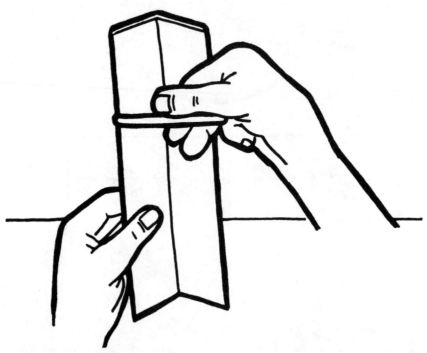

Fig. 3-1. *Folded cardboard and a rubber band makes a support for the thermometer.*

nearby. Have the cardboard side facing the sun so that the light doesn't shine directly on the thermometers. Let them sit for a few minutes until the temperature settles down, then check the readings. The thermometer in the jar should read much higher than the other one. This is because the sun's rays warm the objects they strike. The thermometer

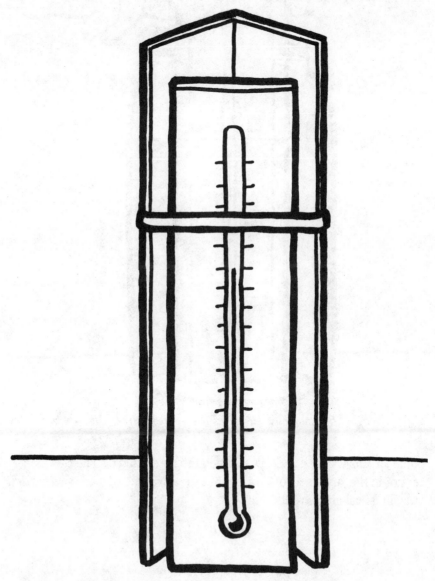

Fig. 3-2. *The thermometer will stand upright inside its support.*

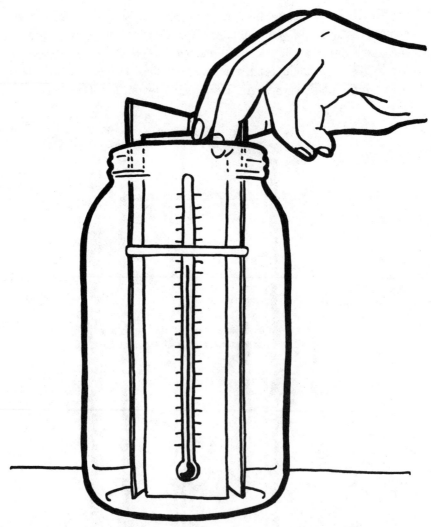

Fig. 3-3. *One thermometer measures the temperature of the air inside the jar.*

in the jar was warmed and this warmed the air around it. The thermometer not in the jar was also warmed, but the surrounding heat was carried away by small air currents.

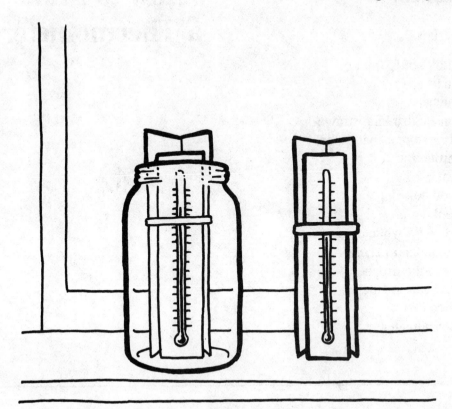

Fig. 3-4. *Both thermometers have their backs to the sun. Notice the higher reading in the jar.*

Experiment 4

How to Make a Thermometer

Materials

- ☐ glass bottle (pint or quart size)
- ☐ pitcher
- ☐ plastic drinking straw
- ☐ cork stopper with a hole through it
- ☐ water
- ☐ food coloring and cooking oil
- ☐ 3″ × 5″ white card
- ☐ candle and matches
- ☐ medicine dropper
- ☐ transparent tape or thread
- ☐ thermometer

In a pitcher, mix a few drops of food coloring in a little more water than you'll need to fill the bottle. Fill the bottle with the colored water and insert the drinking straw an inch or two through the hole of the cork stopper. Press the cork into the opening in the bottle. Have an adult light the candle and drop melted wax around the straw to seal it to the stopper. Tilt the bottle to keep the flame away from the straw. Using the medicine dropper, add more colored water through the open end of the straw. Bring the level up the straw, about an inch above the stopper. Now add a drop of oil on top to prevent the water from evaporating.

Using clear tape or thread, carefully attach the card to the back of the straw. Make a mark on the card alongside the level of the colored water. Compare this mark with a thermometer and then write the temperature on the card. The two thermometers must be at the same location.

Fig. 4-1. *Fill the bottle with colored water.*

Take both thermometers outside for additional readings to mark on the card. This will determine how well your homemade thermometer follows a regular thermometer. The level in the straw will raise or lower because the liquid inside the bottle expands when heated and contracts when cooled.

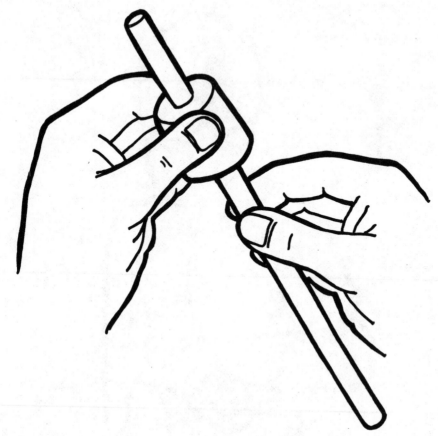

Fig. 4-2. *The straw should stick through the cork an inch or two.*

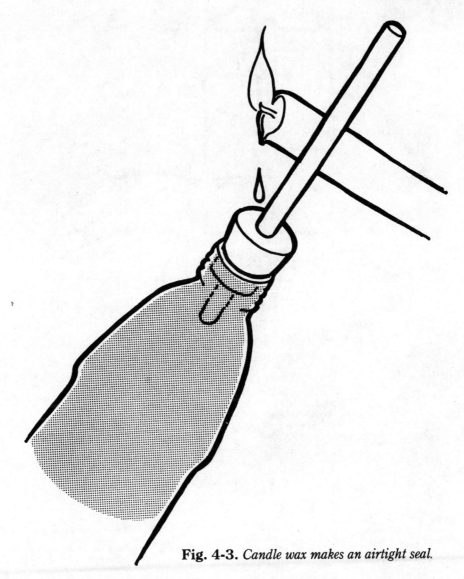

Fig. 4-3. *Candle wax makes an airtight seal.*

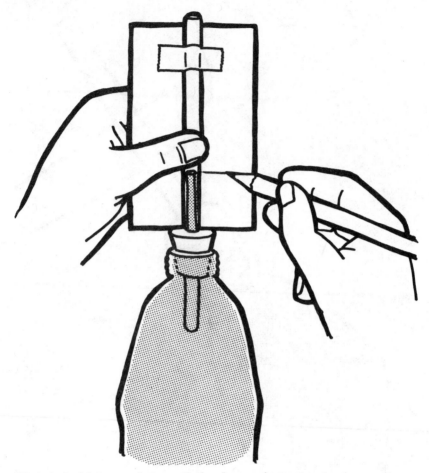

Fig. 4-4. *Make a scale to mark the changes in temperature.*

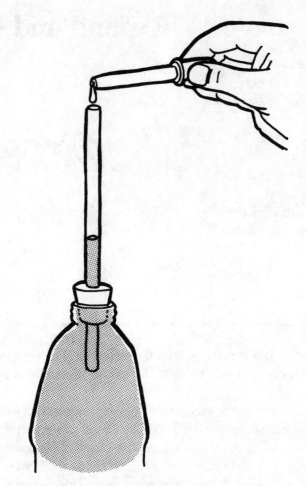

Fig. 4-5. *A drop of cooking oil prevents evaporation.*

Experiment 5

Materials

- [ ] candle and matches
- [ ] flat head screw (do not use coated or galvanized screws)
- [ ] screw eye (the kind found on a screen door) (do not use coated or galvanized screw eyes)
- [ ] 2 wooden broom sticks handles, about one foot long
- [ ] screwdriver

How Metals Expand and Contract

Warning: Do not heat any type of galvanized or coated nails or screws. Heating of any galvanized metal produces a poisonous gas.

The flat head of the screw must be large enough to just barely fit through the screw eye (see Fig. 5-1). Drive the screw part way into the end of one of the wooden handles and the screw eye into the end of the other handle. You might have to use a small file on the head of the screw to make it a close fit through the screw eye. This experiment will be more successful if the screw and the screw eye are made of the same kind of metal.

Have an adult hold the wooden handle and place the head of the screw in the candle flame for several minutes. Then try to fit it through the screw eye. It won't fit because heat caused the molecules in the metal to move faster and farther apart. The metal expanded just enough so that the screw head would not fit through the screw eye.

Cool the screw by holding it under cold running water and try it again. It should pass through the screw eye because, as the metal cooled the molecules slowed down and came closer together. The metal contracted and the head was able to fit through the screw eye.

Heat both the head of the screw and the screw eye. Try the fit again. This time they both expanded and the screw was able to pass through the eye.

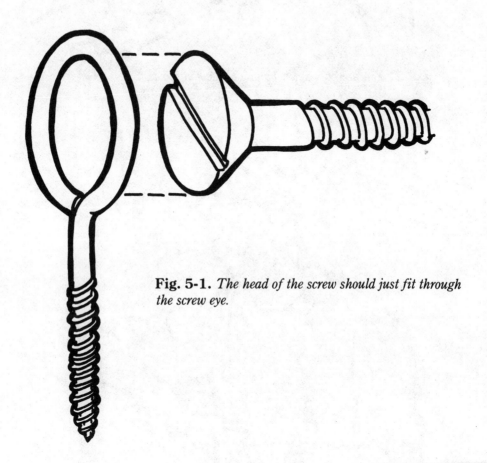

Fig. 5-1. *The head of the screw should just fit through the screw eye.*

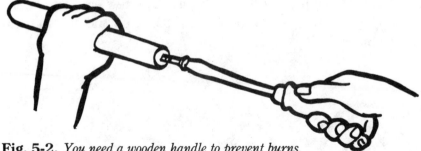

Fig. 5-2. *You need a wooden handle to prevent burns.*

Fig. 5-3. *Heat causes the metal to expand.*

Fig. 5-4. *Both parts expanded at the same rate.*

Experiment 6

Materials

- [ ] thin copper wire (about 3 feet)
- [ ] weight (about $1/2$ lb.) (washer or fishing sinkers)
- [ ] ruler
- [ ] 2, two-liter pop bottles filled part way with sand, gravel or rocks
- [ ] candle and matches

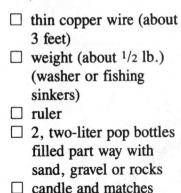

How Copper Wire Expands and Contracts

Stretch the copper wire between the two bottles as shown in Fig. 6-1. Attach the wire at a height that will allow the candle to heat the wire. Fasten the weight near the center of the suspended wire. Using the ruler, measure the height of the wire at the point where the weight is attached.

Have an adult heat the copper wire with the flame of the candle while you monitor the height near the weight. As the wire is heated, it begins to expand and the weight starts to lower. Remove the candle and allow the wire to cool. The wire will contract and the weight will return to near its original height. See Figs. 6-2 through 6-4.

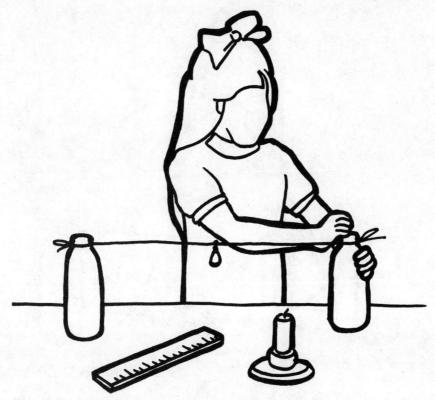

Fig. 6-1. *Suspend the copper wire almost a foot above the surface of the table using two supports.*

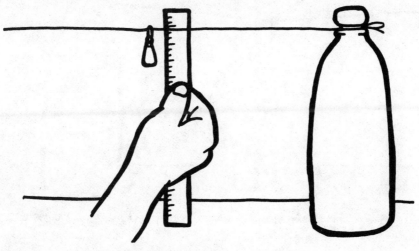

Fig. 6-2. *Try to measure accurately.*

Fig. 6-3. *The heated wire expands.*

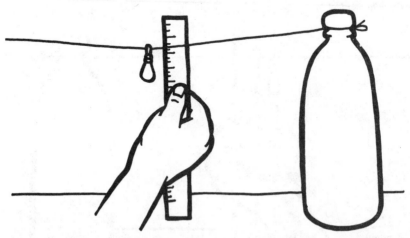

Fig. 6-4. *Compare the measurements to see how far the weight lowered.*

Experiment 7

Materials

- ☐ stiff wire (coat hanger)
- ☐ support with a clamp (a book end)
- ☐ candle and matches
- ☐ coin or cardboard disk

How a Wire Clamp Expands

Bend the wire into the shape of a triangle (see Fig. 7-1) with the ends touching at one of the corners. Clamp the wire triangle to the support so that the triangle is horizontal, with each corner at near the same height (see Fig. 7-2). Suspend a small coin or disk between the points where the wire ends touch.

Carefully light the candle and place it on the opposite side of the wire triangle. The coin will drop. This is because the heated wire expanded causing the two ends to separate.

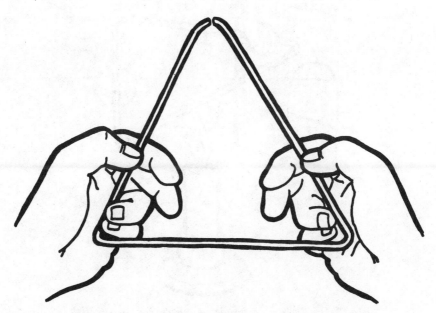

Fig. 7-1. *The opening must be small enough to clamp the coin or disk.*

Fig. 7-2. *A metal bookend makes a good support.*

Fig. 7-3. *The wire expands and drops the coin or disk.*

Experiment 8

Conduction of Heat along Metal

Materials

- ☐ metal rod or large wire (about 12 inches long) (copper, brass or aluminum)
- ☐ 6 or 8 thumb tacks
- ☐ candle & matches
- ☐ protective gloves

Attach the tacks to the metal bar with melted wax from the candle (see Figs. 8-1 and 8-2). Position the tacks about an inch apart beginning at one end of the rod. Be sure to wear protective gloves because hot wax is hot and can burn.

After the wax has cooled a few minutes, place the end of the rod near the first tack over the candle flame. Holding the rod in the same place, notice how the tacks begin to drop from the rod. This shows that heat moves along the rod by conduction. Conduction in physics is a form of energy, such as electricity and heat, that travels along a material from one molecule to another.

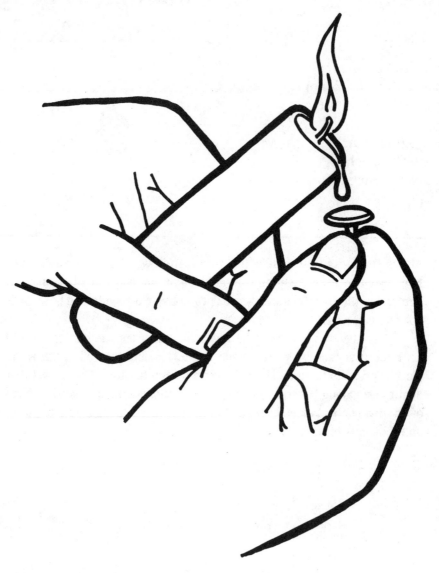

Fig. 8-1. *Use candle wax to stick the tacks to the rod.*

Fig. 8-2. *The tacks should be placed about an inch apart.*

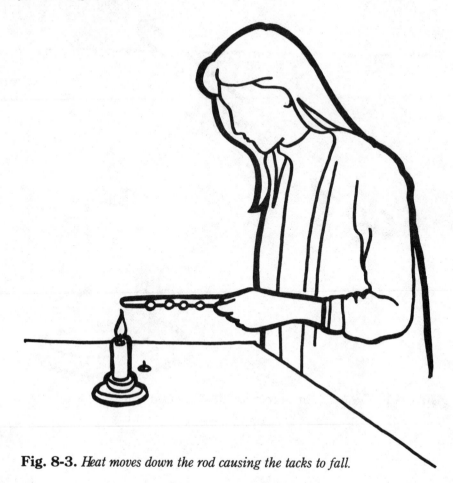

Fig. 8-3. *Heat moves down the rod causing the tacks to fall.*

Experiment 9

Materials

- ☐ coin (quarter or half dollar)
- ☐ old handkerchief
- ☐ small candle or match

Heat Conduction of a Coin

Twist the handkerchief around the coin so that a single layer of the cloth is stretched tightly across the flat side of the coin. Hold this flat side over the flame of the candle for a few seconds. The cloth will not burn.

This is because the metal of the coin is a good conductor and it conducts the heat away from the cloth, keeping the temperature of the cloth below its burning point.

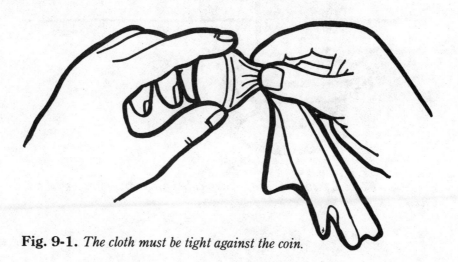

Fig. 9-1. *The cloth must be tight against the coin.*

Fig. 9-2. *The coin keeps the cloth from catching fire because it is a good conductor.*

Experiment 10

Materials

- [ ] $1/2$ inch copper tube (about 6 inches long)
- [ ] strip of paper about ($2^1/_2$" × 6")
- [ ] knitting needle
- [ ] candle and matches

Discovering Kindling Temperature and Heat Conduction

Kindling temperature is the temperature that a material will start to burn. In this experiment, the copper tube will prevent the paper from reaching its kindling temperature.

Place the two ends of the paper together to form a loop. Fasten these ends together by piercing them with the end of a knitting needle (see Fig. 10-1). Place the copper tube inside the paper loop with some

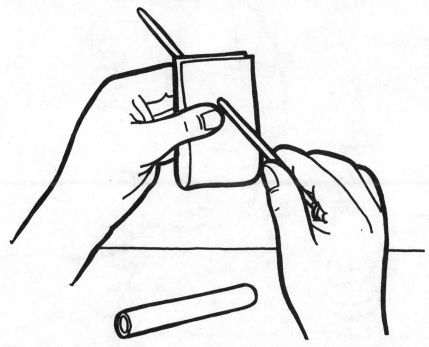

Fig. 10-1. *Suspend a paper loop from a stiff wire or knitting needle.*

of the tube extending past each side of the paper. Hold the other end of the knitting needle and suspend the bottom of the paper loop containing the copper tube over the flame of the candle as shown in Fig. 10-2. The paper will not burn. The copper tube conducts the heat away from the paper, keeping its temperature below the kindling temperature, or the point where the paper will burn.

Fig. 10-2. *The copper tube keeps the paper from burning.*

Experiment 11

Materials

- [ ] 2 wires of different metals, about 12 inches long (copper or aluminum wire and iron coat hanger)
- [ ] 6 metal tacks or paper clips
- [ ] candle and matches
- [ ] protective gloves

Heat Conduction of Different Metals

Twist one end of the wire from the coat hanger and one end of the copper wire together to form one wire nearly two feet long (see Fig. 11-1). Using melted candle wax, fasten three or four tacks about 2 inches apart on each side of the twisted connection. Allow the wax to cool; hold the twisted connection over the candle flame and pay careful attention to when the tacks drop from the two different wires. The tacks attached to the copper wire will start to drop first. This is because the copper molecules conduct heat better than iron molecules. Copper also conducts

Fig. 11-1. *Twist two wires of different metals together.*

electricity better than iron. Some metals conduct heat (and electricity) much better than others. Copper makes an excellent conductor. This is why the bottom of some cooking pots and pans are copper coated.

Fig. 11-2. *Use candle wax to attach the tacks.*

Fig. 11-3. *Heat the center of the two wires and see which tack falls first.*

Experiment 12

Measuring the Expansion of Copper

Materials

- ☐ length of rigid, hollow copper tubing (about 5 feet long)
- ☐ clamp
- ☐ table
- ☐ knitting needle or large nail
- ☐ cardboard pointer (about 12 inches long)
- ☐ hair dryer
- ☐ glue

Place the copper tube lengthwise near the edge of a table. Clamp one end securely to the table and place the knitting needle under the other end as shown in Fig. 12-2. The end of the needle should extend past the edge of the table. Use the glue to fasten the middle of the pointer to the

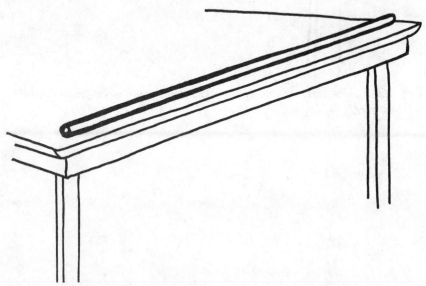

Fig. 12-1. *Place a length of rigid copper tubing on the edge of a table.*

end of the knitting needle. Turn the hair dryer on high heat and blow hot air through the tube from the clamped end. Point the hair dryer so that the air does not directly hit the pointer. Notice the movement of the pointer. The hot air warms the tube causing it to expand. This expansion rolls the needle making the pointer move.

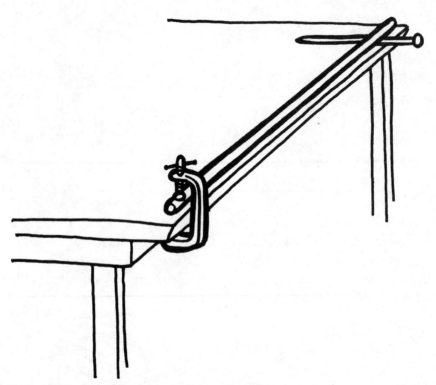

Fig. 12-2. *The clamped end of the rod should apply some pressure on the knitting needle.*

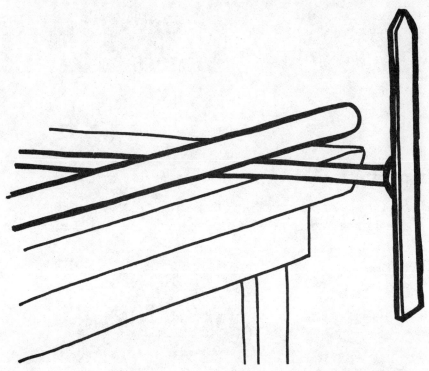

Fig. 12-3. *A pointer is attached at the middle to the knitting needle.*

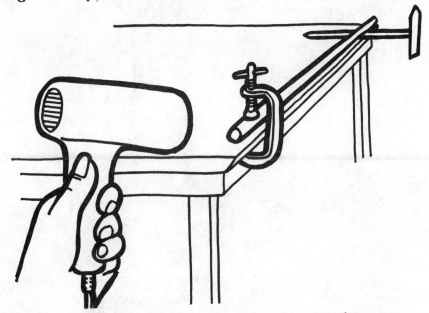

Fig. 12-4. *Heat causes the tube to expand and makes the pointer move.*

Experiment 13

Materials

- ☐ wire screen (old tea strainer)
- ☐ candle and matches
- ☐ piece of paper

Principles of the Davy Lamp or Miners Safety Lamp

Hold the metal screen over the flame of the candle. Notice that the flame does not come through the screen. The heat of the flame is conducted

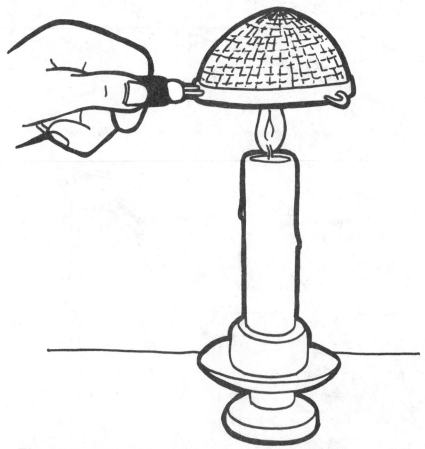

Fig. 13-1. *Hold an old tea strainer over the flame.*

away by the wires. Lower the piece of paper until it is almost touching the screen. It does not burn. The wire screen keeps the paper from reaching its kindling temperature. This gave Sir Humphry Davy the idea of the miners safety lamp in the early eighteenth century, which helped to prevent explosions caused by gases in coal mines. The Davy Lamp was simply a candle enclosed in a wire screen. The screen kept the candle from causing any gas in the mine to reach its kindling temperature.

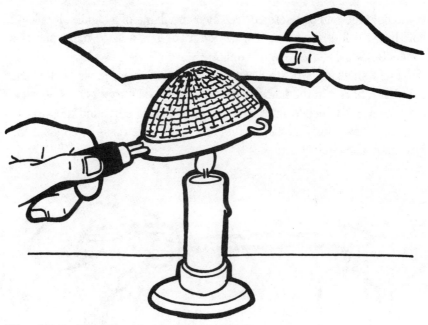

Fig. 13-2. *The flame is contained inside the screen.*

Experiment 14

Materials

- ☐ steel wool
- ☐ pliers
- ☐ candle and matches

Burning Iron (Heat against Mass)

If you hold the end of a screwdriver over the flame of a candle, it will get hot, but it will never burn. The screwdriver has a large mass and this conducts the heat away from the flame.

Pull a small piece of steel wool apart to form a very loose wad. Using the pliers, hold it in the flame of the candle. It will burn. This is because oxygen, which is necessary for burning, is able to surround the strands of steel. These thin strands of steel have very little mass to conduct the heat away and easily burn.

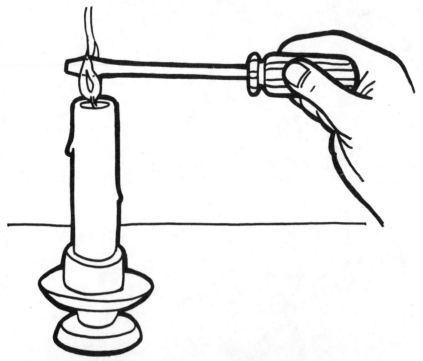

Fig. 14-1. *The screwdriver has too much mass to burn from a small flame.*

Fig. 14-2. *Pull the steel wool apart.*

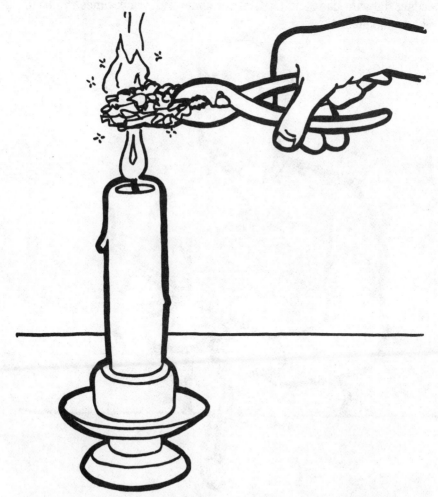

Fig. 14-3. *A small flame is able to ignite the steel wool because breaking it into smaller pieces reduces its mass.*

Experiment 15

Why Bending Wire Becomes Hot

Materials

☐ length of coat hanger wire (about 12″ long)

Grasp the wire so that you can make a bend in the middle and begin bending the wire back and forth. After about five or six times the bend will become hot.

Fig. 15-1. *You can bend a wire coat hanger slowly without generating much heat.*

When the wire is bent, the molecules in the metal are rubbed against each other. The friction of this rubbing generates the heat.

Fig. 15-2. *Rapidly bending the coat hanger will generate much more heat.*

Experiment 16

How a Hammer Generates Heat

Materials

- ☐ hammer
- ☐ large nail
- ☐ heavy iron base (sledge hammer, vise, or anvil)
- ☐ safety goggles

Notice the temperature of the nail. It should feel cool to the touch. Put on safety goggles. Place the pointed end of the nail on the iron base and strike it several times with the hammer. The nail will begin to flatten. Feel the nail again. It will also be warmer. When the hammer hits the nail, the nail begins to change shape. The molecules are forced to move rapidly, generating heat.

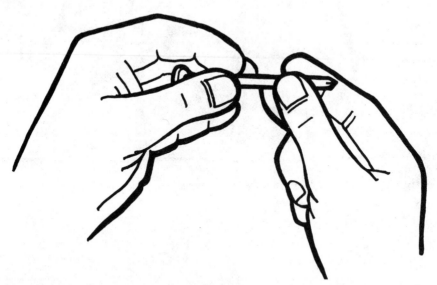

Fig. 16-1. *A nail feels cool because it takes heat from our fingers.*

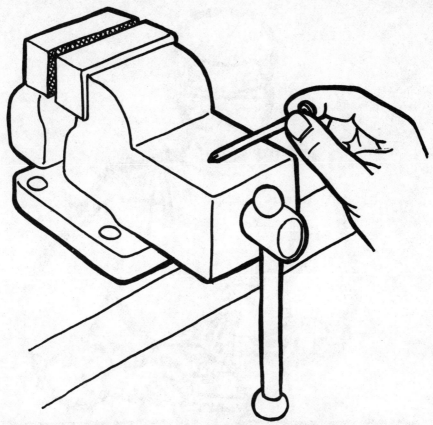

Fig. 16-2. *Place the nail on a solid iron base such as a vice.*

Fig. 16-3. *Energy from the hammer reshapes the end of the nail and causes the iron molecules to move rapidly.*

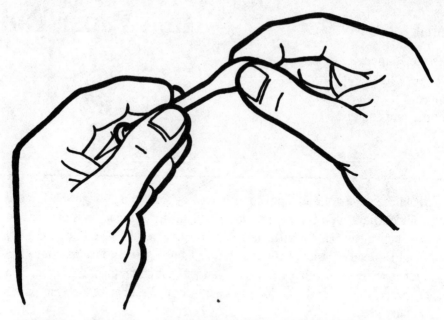

Fig. 16-4. *The flattened end of the nail will be warm and might even be hot.*

Experiment 17

Heating Water in a Paper Pan

Materials

- ☐ sheet of paper
- ☐ 4 paper clips
- ☐ water
- ☐ candle and matches

Fold the edges of the paper to form sides about an inch high. Fasten the four corners with the paper clips. Pour about one-half inch of water in the paper pan. Using both hands and being careful not to burn yourself, hold the pan of water over the candle flame. After a few minutes, the water will be warm but the paper pan will not catch fire. The water is a good conductor and was able to carry away the heat before the paper could ignite.

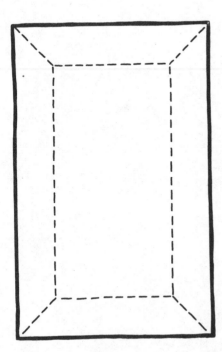

Fig. 17-1. *Make the pan by folding a piece of paper.*

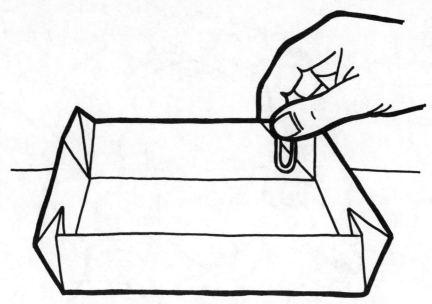

Fig. 17-2. *Hold the corners in place with paper clips.*

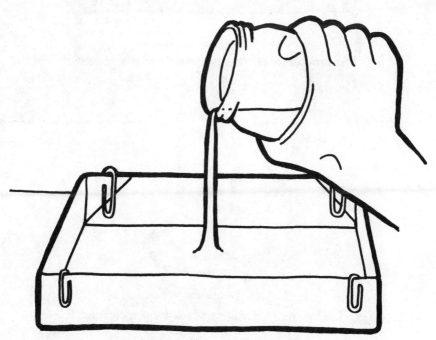

Fig. 17-3. *Pour about a half-inch of water in the pan.*

Fig. 17-4. *You can heat the water without the pan burning.*

Experiment 18

Materials

- [ ] fork
- [ ] spoon
- [ ] two wooden matches
- [ ] drinking glass

Heat Conduction with a Fork and Spoon

Fasten the spoon and fork together by pressing the round end of the spoon into the points of the fork (see Fig. 18-1). Push one of the matches through the top opening of the fork and balance them on the edge of the glass as shown in Figs. 18-2 and 18-3. One end of the match should extend past the edge of the glass, and the other end should stick through the fork a little.

Light the end of the match using the other match. The wood will burn until the fire reaches the fork and the edge of the glass, then it will go out. The fork and the spoon will stay balanced. The wood burns as long as it has enough heat. But when the fire gets to the fork and the glass, they conduct enough of the heat away from the flame that the wood is unable to burn.

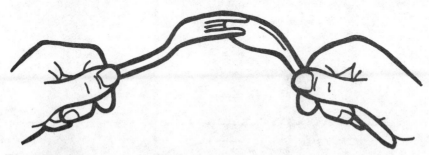

Fig. 18-1. *Fit a fork and spoon of about the same weight together.*

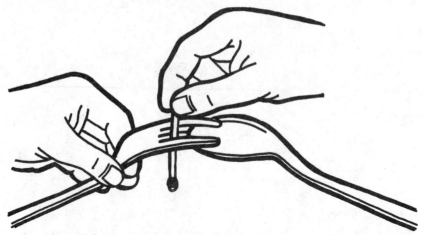

Fig. 18-2. *Press a wooden match through the fork with the tip pointing down.*

Fig. 18-3. *Balance the fork and spoon on the edge of a glass.*

Fig. 18-4. *Light both ends of the match. They will stay balanced.*

Experiment 19

Making Invisible Ink

Materials

- ☐ paper
- ☐ wire screen
- ☐ candle and matches
- ☐ lemon juice or vinegar,
  milk, or grapefruit juice
- ☐ toothpick

The lemon juice, or one of the other liquids, will be the ink. Using the toothpick for a pen, write a message on the paper. Use plenty of ink. When the ink dries the message will disappear.

Fig. 19-1. *Write a message with invisible ink on a sheet of paper.*

To get the message back, place the paper on top of the wire screen and heat them over the flame of the candle. Move the paper around a little to heat it evenly. Keep it high enough so that it doesn't catch fire.

The message will reappear because the liquid used for the ink has a lower kindling temperature than the paper. The ink charred and turned dark when heat was applied.

Fig. 19-2. *After it is dry, heat the paper to see the message.*

Experiment 20

Materials

☐ thermometer
☐ blanket

Measuring Heat from Your Body

Notice the reading of the thermometer, then wrap it in the blanket. Leave it in the blanket a few minutes then check the reading again. The temperature should be the same. Now wrap the thermometer and yourself in the blanket. Leave yourself wrapped a few minutes. Compare the reading. The temperature was higher. The blanket did not produce any heat. It only helped contain the heat that was produced by your body.

Fig. 20-1. *Measure the temperature of the room.*

Fig. 20-2. *Wrap the thermometer in a blanket.*

Fig. 20-3. *See if the temperature remains the same, it should.*

Fig. 20-4. *Measure the temperature inside the blanket when it's wrapped around someone. The blanket alone does not generate any heat, it only contains it.*

Experiment 21

Materials

- ☐ 2 thermometers
- ☐ large jar of water
- ☐ large jar of dirt

Heat and the Breeze at the Beach

Put one of the thermometers in the jar of water and insert the other one into the other jar. Place the two jars in sunlight and monitor the temperatures. The temperature of the dirt will start to rise because the soil will take in heat quicker than the water. At night it looses heat faster.

Fig. 21-1. *Gather a jar of dirt, a jar of water, and two thermometers.*

Near the beach during the day, the ground warms the air causing it to rise and the cooler air from the sea comes in. At night the ground is cooler than the seawater and the air over the water rises while the air over the ground flows out.

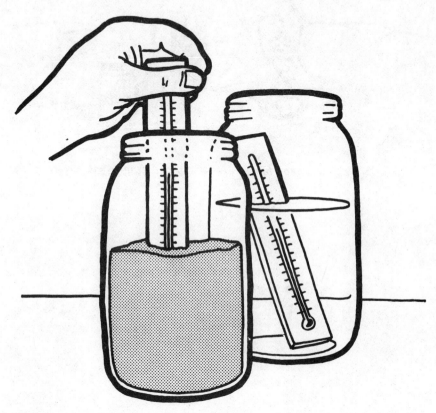

Fig. 21-2. *Place a thermometer in each jar.*

Fig. 21-3. *The different temperatures of soil and water create breezes at the beach.*

Experiment 22

Making a Heat Motor

Materials

- ☐ sheet of paper
- ☐ scissors
- ☐ metal thimble
- ☐ needle
- ☐ spool (sewing thread)
- ☐ wooden pencil with eraser
- ☐ table lamp

Cut the paper into a coil or spiral. Leave enough space in the center to partially insert the thimble (see Figs. 22-1 and 22-2). Make the turns about an inch wide. Make a hole in the center and press the bottom of the thimble part way through the hole. Insert the needle upside down into the eraser. Remove the threaded nut from the top of the lamp shade and place the spool over the threaded stud. Place the pointed end of the pencil into the hole in the spool. Carefully set the thimble in the spiral over the point of the needle. Turn on the lamp and the spiral will begin to turn.

The point of the needle makes very little contact with the thimble and this makes a very good pivot point with little friction.

As the lamp begins to heat the air, the warmer air becomes less dense than the cooler air surrounding it, and so it begins to rise. The warm air pushes on the spiral and it begins to turn.

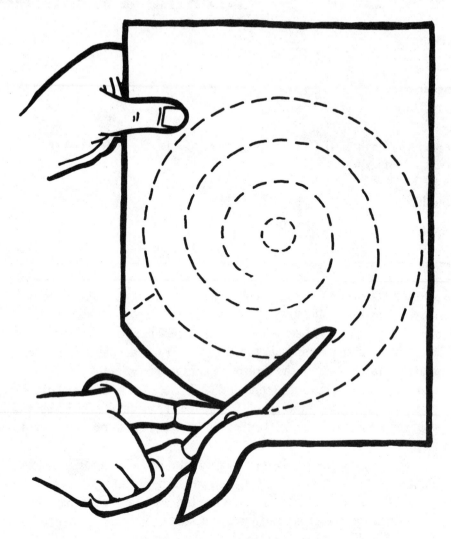

Fig. 22-1. *Cut a spiral from a sheet of paper.*

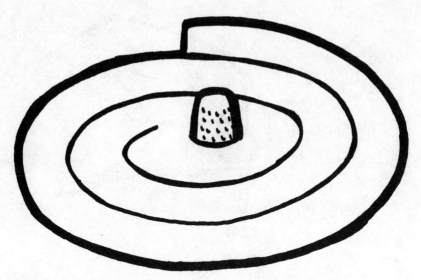

Fig. 22-2. *Place a metal thimble in the center.*

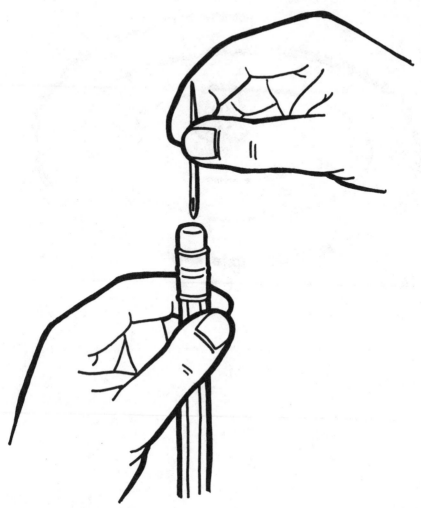

Fig. 22-3. *Push a needle into the eraser with the point facing up. This makes an excellent pivot point.*

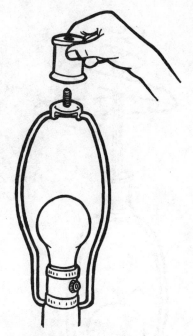

Fig. 22-4. *Use a wooden spool for the base.*

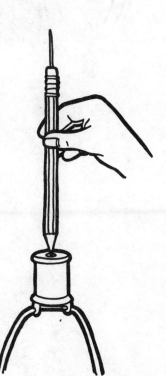

Fig. 22-5. *The pencil and needle form the stand.*

Fig. 22-6. *Heat from the lamp will cause the spiral to turn.*

Experiment 23

Materials

- ☐ glass jug with screw-on cap
- ☐ hot water

How Heat Makes Air Expand

Pour about one inch of hot water into the jug and quickly tighten the cap. Shake the jug vigorously then slowly loosen the cap. You will hear pressurized air escaping from the jug. The air in the jug was cooler when the hot water was poured in. Then when the jug was shaken, the hot water heated the air which expanded and built up pressure inside the jug.

Fig. 23-1. *Run a little, very hot water into a jug.*

Fig. 23-2. *Tighten the cap and shake up the water inside the jug.*

Fig. 23-3. *Gradually loosen the cap. A little air should hiss from the jug.*

Experiment 24

How Air Contracts When Cooled

Materials

- ☐ small balloon
- ☐ tape measure
- ☐ two drinking glasses
- ☐ running water (hot and cold)

Blow up the balloon to a diameter of about 6 or 8 inches. Tie the opening securely. Heat both glasses by holding them under hot running water. Remove the glasses from the hot water and turn on a stream of cold water. Quickly place the openings of the heated glasses to each side of

Fig. 24-1. *Blow up a small balloon.*

the balloon and cool them under the faucet. A portion of the balloon will be pulled inside each glass, attaching them to the balloon. The connection may be strong enough to allow one glass to be lifted by the other. The air in each glass was first heated then trapped inside by the sides of the balloon. Then the cold water cooled this air and it contracted sealing the glasses to the balloon.

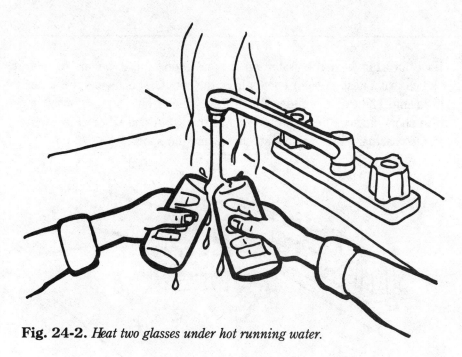

Fig. 24-2. *Heat two glasses under hot running water.*

Fig. 24-3. *When the glasses are cooled they will stick to the balloon.*

Experiment 25

Materials

- ☐ tea kettle
- ☐ water
- ☐ stove

How Heat Creates Steam

Have an adult heat the water on the stovetop until it begins to boil. Notice that a mist of vapor comes from the spout. This means the water is boiling. The vapor is often thought of as steam, but steam is invisible. The short distance between the end of the spout and where the vapor forms is actually the invisible steam leaving the spout.

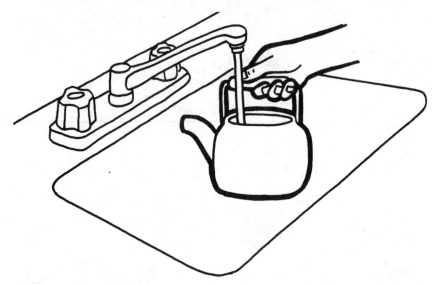

Fig. 25-1. *Fill a tea kettle about half full of water.*

Fig. 25-2. *The steam is the invisible area between the kettle and the vapor.*

Experiment 26

Why Popcorn Pops

Materials

- ☐ skillet with cover or pop-corn popper
- ☐ popcorn
- ☐ stove

Have an adult heat the corn until it begins to pop. You will soon hear miniature explosions. This is because there is moisture inside each kernel of corn. This moisture is contained in a hard, airtight shell. The heat causes this moisture to expand creating pressure. This pressure explodes the corn.

Fig. 26-1. *Popcorn contains a small amount of moisture inside each kernel, which heats and expands, causing the kernels to explode.*

Experiment 27

Heat and Atmospheric Pressure

Materials

- [ ] clean, metal can with a screw-on cap (see Fig. 27-1)
- [ ] water
- [ ] stove
- [ ] 2 pieces of cloth or gloves
- [ ] kitchen sink or large pan

Have an adult pour about one-half inch of water into the can and heat it on the stove. DO NOT put the cap on. DO NOT use a dirty can. Let the water come to a boil with vapor coming from the opening of the can. Use gloves (the can will be very hot) to remove the can from the stove and screw the cap on tight. DO NOT put the cap on until the can is away from the stove. Place the can in the sink and run a little cold water over it. The can will instantly be crushed. After the can has cooled, remove the cap and pour out the water. With some hard blowing, you can probably blow up the can, like you would a balloon, to near its original shape. The can was crushed because the boiling water pushed out most of the air from the can. When the cap was screwed on, the hot moisturized air was trapped inside. When the cold water began to cool the air in the can, most of it turned back into water and this greatly reduced the pressure. On the outside, the normal air pressure (about 14.7 pounds per square inch at sea level) could have pressed more than a ton of pressure, easily crushing the can.

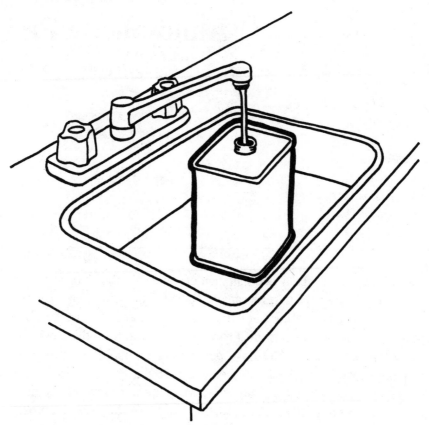

Fig. 27-1. *Pour a small amount of water in a tin can.*

Fig. 27-2. *Bring the water to a boil.*

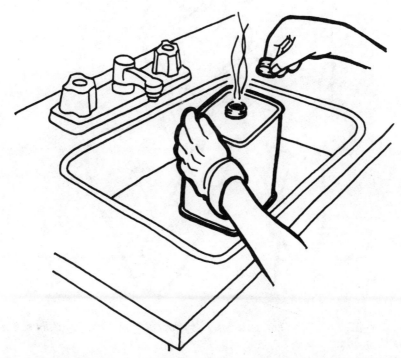

Fig. 27-3. *Wearing protective oven mitts or gloves, take the can to the sink and tighten the cap.*

Fig. 27-4. *Cool the can with cold water. The can will instantly collapse.*

Fig. 27-5. *Empty the hot water and you probably can blow out the can to near its original shape.*

Experiment 28

Boiling Water with Water

Materials

- ☐ pot
- ☐ water
- ☐ stove
- ☐ small bottle
- ☐ tongs

Fill the pot half full of water and and have an adult heat it on the stovetop until it starts to boil. Fill the small bottle about half full of water, and using the tongs, suspend it in the boiling water without it touching the bottom of the pot. After several minutes, the water in the bottle will still not boil. Water boils at 212 degrees Fahrenheit. It can bring the water in the bottle up to that temperature, but at that point, the boiling water is unable to supply the little extra heat needed to make the water in the bottle boil. Adding several spoons of salt to the boiling water will raise its boiling point enough to make the water in the bottle boil.

Fig. 28-1. *Heat from a stove can easily boil water.*

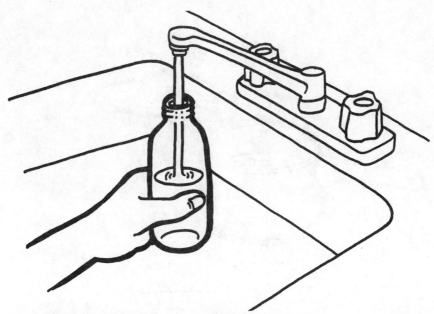

Fig. 28-2. *Add some water to a small bottle.*

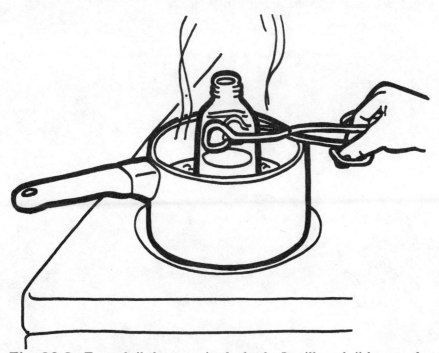

Fig. 28-3. *Try to boil the water in the bottle. It will not boil because the water in the pan does not have enough heat.*

Fig. 28-4. *Salt raises the boiling point of water.*

Experiment 29

Latent Heat

Materials

☐ drinking glass
☐ several ice cubes or crushed ice
☐ thermometer

Place the thermometer in the glass and fill the glass about half full of ice. Put it on a table and monitor the temperature changes as the ice melts. The temperature will drop to almost 32 degrees Fahrenheit. At that point, the ice will melt. Notice that the temperature of the water will also be at, or near, 32 degrees. The water will remain at 32 degrees until the ice is almost gone, then its temperature will start to rise.

Ice molecules are arranged in crystals. As the warmth of the room melts the ice, all of the heat is used to tear the crystals apart. This heat is called latent heat or heat of fusion. After the ice is melted, the temperature of the water will start to rise. If the water was in a container that can be heated, and enough heat was added, the temperature would increase until it reaches 212 degrees Fahrenheit, which is the boiling point of water. At this point, the water is changed into steam. This heat is called the heat of vaporization.

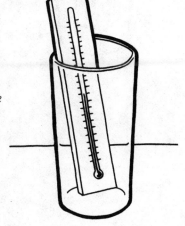

Fig. 29-1. *Measure the temperature inside an empty glass.*

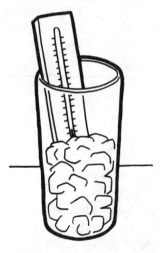

Fig. 29-2. *Fill the glass with ice and watch the temperature. It should drop to about 32 degrees.*

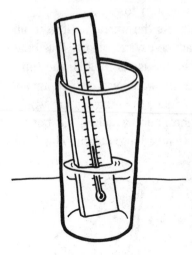

Fig. 29-3. *The water will stay at 32 degrees until the ice has melted.*

Experiment 30

Materials

- [ ] quart jar
- [ ] drinking cup
- [ ] India ink (waterproof ink)
- [ ] hot and cold water
- [ ] sheet of white paper

Transferring Heat through Liquids by Convection Currents

Fill the jar about half full of cold water from the faucet and set it aside to settle the water. Fill the cup nearly full of hot water and sit it on a table. Drop one drop of ink in the center of the cold water and place the jar on top of the cup of hot water (see Figs. 30-3 and 30-4). The drop of ink will settle to the bottom of the jar. Stand the sheet of paper behind the jar to make the movement easier to see. The ink will start to rise from the bottom forming slow swirls. The heat from the water in the cup warmed the water in the bottom of the jar. The warm water is less dense and rises. This causes the cooler, heavier water to flow down to the source of heat. This creates a circular motion called a convection cycle which is also sometimes called a convection current.

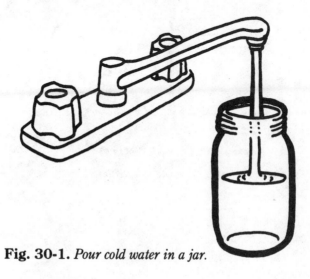

Fig. 30-1. *Pour cold water in a jar.*

Fig. 30-2. *Fill a cup with hot water.*

Fig. 30-3. *Put a drop of ink in the cold water.*

Fig. 30-4. *Heat will cause currents to flow in the cold water.*

Experiment 31

Expansion and Contraction of Water and Ice

Materials

☐ quart jar
☐ ice
☐ water

Place the jar on a level table or counter top and fill it about half full of ice. Slowly add water until the jar is completely full with the ice pressed below the surface. Allow the ice to float free. The water level will drop some. After several minutes, the ice melts and the water is slightly lower still. As the ice finishes melting, the water level should be near its original level, but as the water warms to room temperature, the level rises slightly. This is because the water was cooled and contracted as the ice melted. Then as the water warmed it expanded.

Fig. 31-1. *Fill a jar about half full of ice.*

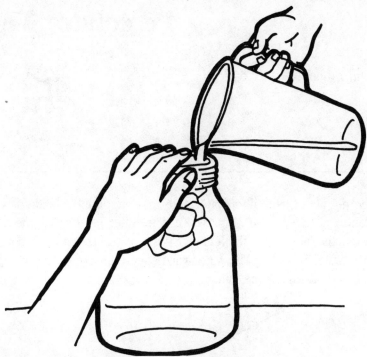

Fig. 31-2. *Fill the jar completely full of water.*

Fig. 31-3. *Monitor the different water levels as the ice melts and the water rises to room temperature.*

Experiment 32

Materials

- ☐ metal pot
- ☐ small metal funnel
- ☐ water
- ☐ stove

How a Percolator Works

Fill the pot about half full of water and place the funnel upside down in the center of the pot (see Fig. 32-2). The water level should be about three-fourths the way up the funnel. Sit the pot on the stove and slowly heat the water to a boil. The water will begin to bubble out of the opening in the small end of the funnel.

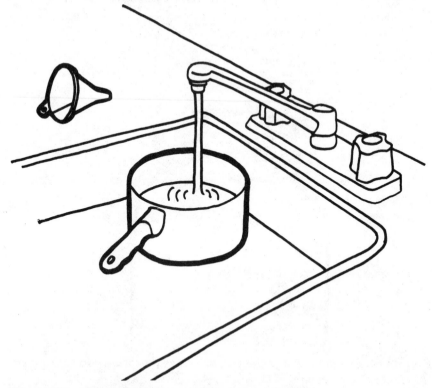

Fig. 32-1. *Fill a pot with two or three inches of water.*

The heat from the stove has expanded the water molecules and changed the water inside the funnel to a mixture of steam bubbles and very hot water. This mixture is lighter and less dense than the water outside the funnel. The cooler, heavier water on the outside is forced down (by gravity and air pressure in the room) inside the funnel and pushes the lighter steam and water up the funnel to percolate out of the opening.

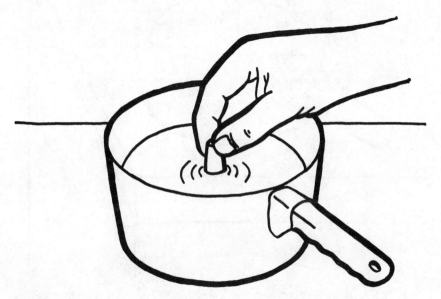

Fig. 32-2. *Place a metal funnel upside down in the pot.*

Fig. 32-3. *Hot water percolates out of the funnel.*

Experiment 33

Why Boiled Water Freezes Faster

Materials

- ☐ 3 glasses, the same size
- ☐ boiled water
- ☐ tap water from faucet
- ☐ saltwater mixture
- ☐ freezer

Have an adult boil some water for about two minutes and allow it to cool. Fill one glass nearly full of the boiling water; fill the second glass to the same level with water from a faucet; the third glass should be filled to the same level with the saltwater.

Allow all three glasses of water to come to room temperature, then place them in a freezer. The boiled water in the first glass will begin to freeze first. The plain water from the faucet will freeze next. The salt

Fig. 33-1. *Assemble some cooled water that has been boiled, tap water, and some salt water.*

water will freeze last. Most anything added to water makes it take longer to freeze or lowers its freezing point. Air is one of these things. Boiling water causes most of its air to be released, so it was the first to freeze. Tap water has some air in it. It froze next. Salt lowers the freezing point of water even more, causing it to freeze last.

Fig. 33-2. *Place them in a freezer to determine which one freezes first.*

Experiment 34

Changing the Melting Point of Ice

Materials

☐ 2 ice cubes
☐ paper towel

Place the two ice cubes on a paper towel so that they are touching each other. After a few seconds, see if they are stuck together. They shouldn't be. Place one cube on top of the other and press hard. Both surfaces should be flat together. Release the pressure and the cubes will be stuck together. The pressure lowered the melting point of the ice and a small layer of water formed between the cubes. Then when you released the pressure, the water froze again and the cubes were stuck.

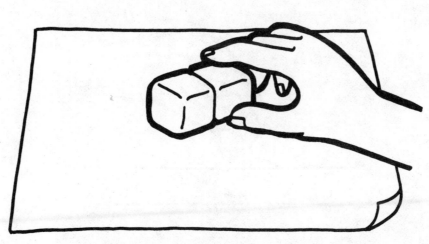

Fig. 34-1. *Place one ice cube against another to see if they will stick together.*

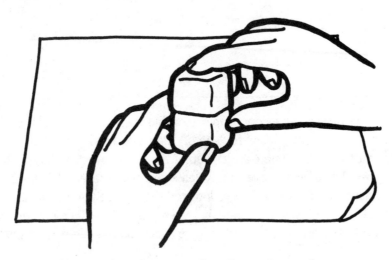

Fig. 34-2. *Place one ice cube on top of another and press down.*

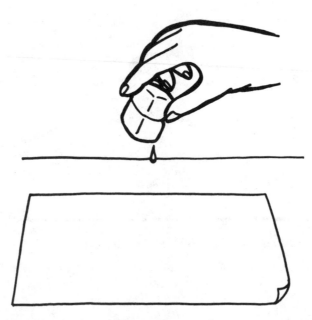

Fig. 34-3. *The two ice cubes are frozen together.*

Experiment 35

Materials

- [ ] length of thin wire from a lamp cord
- [ ] ice cube
- [ ] pot of water
- [ ] 2 small pieces of aluminum foil
- [ ] kitchen sink
- [ ] empty pop can
- [ ] fork
- [ ] string or tape
- [ ] wire stripper or pliers

Pulling a Wire through an Ice Cube

Begin by striping the insulation from an old lamp cord with wire strippers or a pocket knife. **Be extremely careful.** A lamp cord is made of many strands of thin copper wire. You'll need about twelve inches of one strand. See Figs. 35-1, 35-2, and 35-3.

Securely fasten the handle of the fork to the top part of the handle of the pot. This can be done with string or tape. The points of the fork will be level and pointing away from the pot as shown in Fig. 35-4. Mount the fork so that the pointed end will form a cradle for the ice cube. Now fill the pot with water and place it next to the sink so that the fork extends over the sink (see Fig. 35-5).

Bend the tab of the pop can up to make a lifting point and fill the can with water as shown in Fig. 35-6. Feed one end of the wire through the opening in the tab of the can (see Fig. 35-7). Form a loop about five or six inches long and tie the wire in a knot. Lift the can of water by the wire and fit the wires through the center openings of the fork as shown in Fig. 35-8. Place the ice cube through the loop and over the fork. Lay a piece of aluminum foil between the ice and the fork on each side of the wire loop (see Fig. 35-9). This will slow the melting of the ice. Now lower the wire so that the weight of the can is suspended by the ice cube. In a few minutes the wire will begin to move through the ice. After about twenty or thirty minutes the can will fall and the wire will move

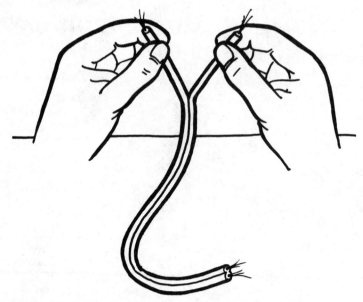

Fig. 35-1. *Separate the two halves of an old lamp cord.*

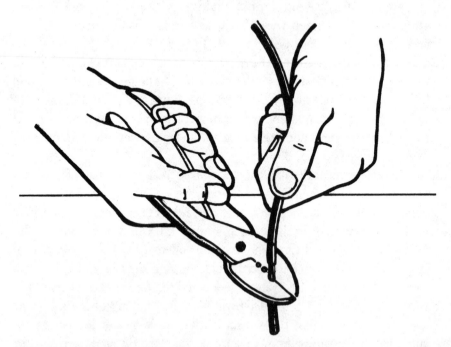

Fig. 35-2. *Cut off a piece of wire about 12 inches long.*

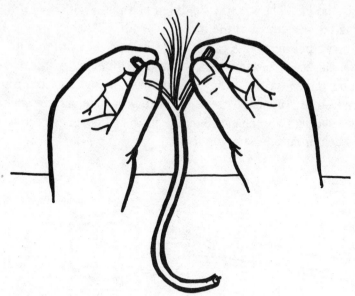

Fig. 35-3. *Peel the insulation away to get to the strands of wire.*

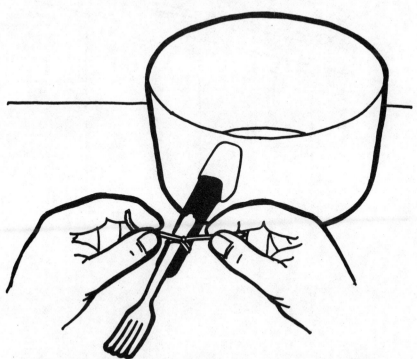

Fig. 35-4. *Use string or tape to attach the fork to the handle. The fork prongs should face away from the pot.*

through the ice. The ice will still be in one piece with only a thin line showing the movement of the wire.

Gravity pulling on the can of water caused the wire to apply pressure on the ice. This pressure lowered the melting point of the ice, which melted and allowed the wire to move through the ice. After the wire moved, the pressure above the wire was gone and the melted ice refroze, sealing the opening. Because it was a copper wire it also conducted heat from the air in the room.

Fig. 35-5. *Fill the pot with water so it won't tip.*

Fig. 35-6. *The pop can full of water provides the necessary weight.*

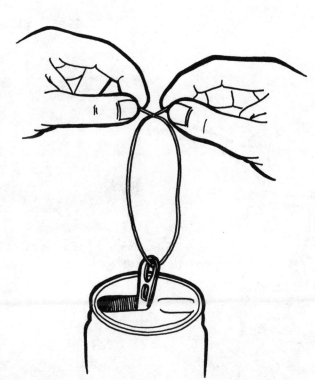

Fig. 35-7. *Make a loop from a strand of copper wire.*

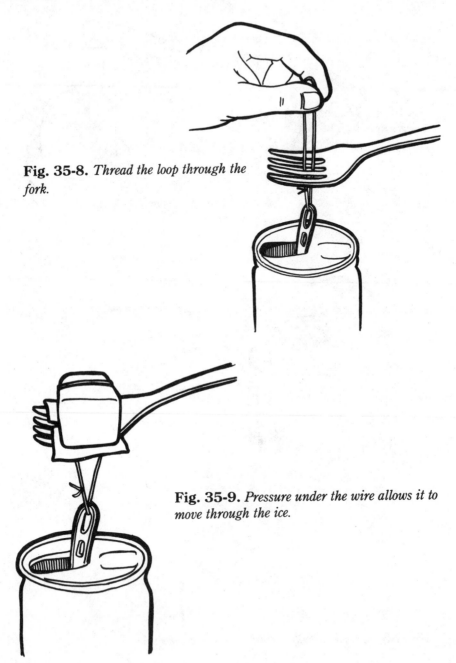

Fig. 35-8. *Thread the loop through the fork.*

Fig. 35-9. *Pressure under the wire allows it to move through the ice.*

Experiment 36

Insulating Abilities of Wet and Dry Materials

Materials

- ☐ one pot holder
- ☐ hot pan

Using a pot holder, pick up the hot pan. Moisten the pot holder and carefully try to pick up the pan again. In the first try, little heat was felt in the hand, but in the second attempt, considerable heat was felt. If the pot holder was completely wet, the hands can easily be burned. It is important to always use dry materials to handle anything hot.

The dry pot holder is made up of fibers with many insulating air spaces. Air is not a good conductor of heat. If the pot holder is wet, these air spaces are filled with water and water is a much better conductor of heat. If the pan is very hot, this heat can turn the water into steam, which can travel through the wet fabric to burn the hands.

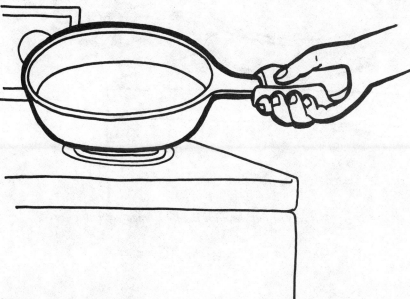

Fig. 36-1. *A dry pot holder provides good insulation against heat.*

Fig. 36-2. *Apply a small amount of moisture to one of the pot holders.*

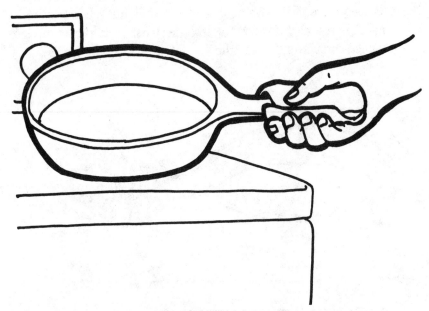

Fig. 36-3. *Heat can be felt through the moistened pot holder.*

Experiment 37

Materials

- ☐ jar
- ☐ candle and matches
- ☐ pan
- ☐ water

How Air Pressure Affects Water Level

Using the wax from a burning candle, attach the candle to the center of the pan (see Fig. 37-1). Be careful not to burn your hands with hot wax. Fill the pan about three-fourths full of water and place the jar upside down over the burning candle as shown in Figs. 37-2 and 37-3. In a short while, the candle will go out, and the water will rise inside the jar. It may even flood the candle.

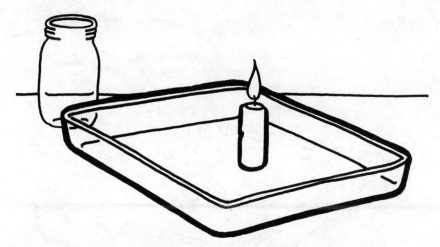

Fig. 37-1. *Place a burning candle in the bottom of the pan. Seal it in place with a little hot wax.*

Fig. 37-2. *Pour in some water.*

Fig. 37-3. *Place the jar over the candle.*

It is often thought that the water rises to fill the space occupied by the oxygen that was burned by the flame of the candle. This is not true. The flame released an equal amount of gases and vapors, mostly carbon dioxide and water, to take the place of the burned oxygen.

What really caused the water to rise, was the heated air trapped under the jar expanded and then bubbled out through the mouth of the submerged jar. As the flame grows dimmer, the air inside the jar cools and contracts. The normal air pressure in the room and outside the jar forces the water up inside the jar.

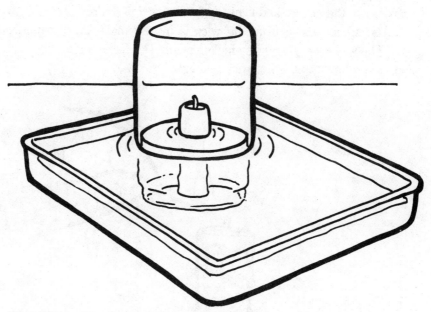

Fig. 37-4. *The candle goes out and water rises in the jar.*

Experiment 38

Materials

- ☐ candle and matches
- ☐ protective gloves

Lighting a Candle Through Its Smoke

Light the candle and let it burn until the wax around the wick has melted some. Blow out the flame. Notice the smoke rising from the wick. Light the match and hold its flame in the rising smoke. The flame will quickly move down the smoke and light the candle.

The smoke rising from the wick is mostly made up of vaporized wax. The wax is very near its kindling point. The flame ignites the vapor and travels down to the hot wick and lights it.

Fig. 38-1. *Light a candle and let it burn a little.*

Fig. 38-2. *Blow out the candle. Smoke will rise from the wick.*

Fig. 38-3. *Flame travels down the smoke to the wick.*

Experiment 39

How Carbon Dioxide Prevents Flame

Materials

- [ ] large bowl or jar
- [ ] small glass
- [ ] two candles of different heights and matches
- [ ] vinegar
- [ ] baking soda

Mount a short candle and one a little taller in the bottom of the bowl as shown in Fig. 39-1. The top of the candles must be below the rim of the bowl. Pour vinegar into the glass and place it in the bowl. Light the candles and, while they are burning, drop a spoonful of baking soda into the glass of vinegar. This will foam and come out of the glass. After a few seconds, the shortest candle will go out. Then the taller candle will be extinguished.

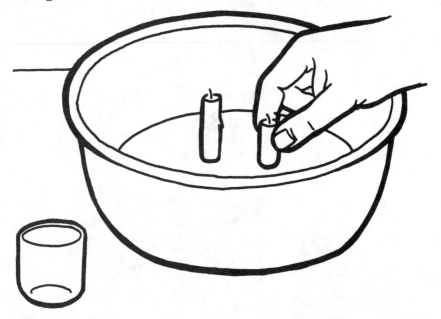

Fig. 39-1. *Place the two candles in the bowl. The top of the candles must be below the rim of the bowl.*

When baking soda is added to vinegar, it produces carbon dioxide gas. This is about one and a half times heavier than air. So it fills up the bowl from the bottom. It also cuts off the supply of oxygen to the flame. As the level of carbon dioxide reaches each candle it puts out the flame.

Fig. 39-2. *Add a small amount of vinegar in a glass.*

Fig. 39-3. *Add a spoonful of baking soda to the vinegar.*

Fig. 39-4. *Carbon dioxide gas builds up in the bottom of the bowl.*

Experiment 40

Materials

☐ candle and matches
☐ piece of aluminum foil

How a Cooled Flame Can Deposit Carbon

Hold one end of the aluminum foil in the flame of the candle. In seconds, it will be covered with black soot, which is made up mostly of carbon particles. This black soot formed because the flame was cooled by the aluminum foil. Normally, a flame produces carbon and hydrogen, which form carbon dioxide gas and water vapor. These gases usually mix with the surrounding air and disappear. The heat of the flame is produced when the oxygen is mixed with the carbon and hydrogen. Once the flame was cooled by the aluminum foil, the carbon was unable to mix with the oxygen at the lower temperature. This formed the soot (mostly carbon) that was deposited on the aluminum.

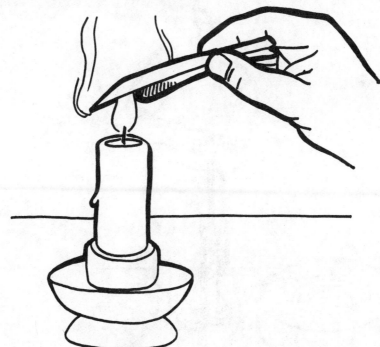

Fig. 40-1. *Aluminum foil lowers the temperature of the flame.*

Experiment 41

Materials

☐ tire pump
☐ flat tire or basketball

Why a Tire Pump Generates Heat

Operate the pump to inflate the tire. After several minutes of hard pumping, feel the lower part of the barrel of the pump. It should feel quite warm. This heat is produced by friction. The bottom part of the pump is heated because the molecules making up the air are pressed closer together and they are forced to rub against, and strike each other, more than normal. This rubbing and striking of the molecules is what produces the heat. There is also some added heat caused by the friction of the piston rubbing against the inside of the barrel.

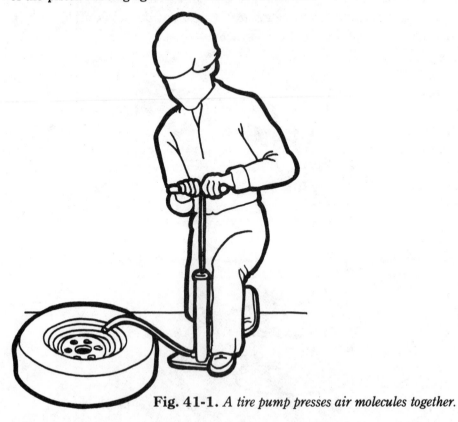

Fig. 41-1. *A tire pump presses air molecules together.*

Fig. 41-2. *Air molecules rubbing against one another generates heat.*

Experiment 42

Materials

☐ electric lamp (desk lamp)

Heat Transferred by Radiation

With the lamp off, hold your hand, palm up, two or three inches below the bulb. Now turn on the lamp. You should feel the heat almost as soon as the bulb lights. The heat travels by radiation almost instantly. When heat travels by radiation it can even travel across a vacuum. Heat from the sun warms the earth by radiation across the vacuum in space.

Fig. 42-1. *Hold one hand beneath a lamp.*

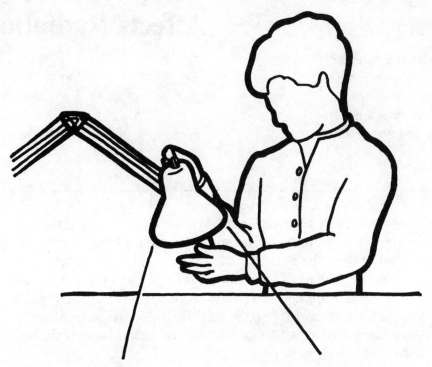

Fig. 42-2. *Heat from the lamp travels to your hand by radiation.*

Experiment 43

Materials

- [ ] candle and matches
- [ ] two shiny tin cans, the same size
- [ ] two thermometers
- [ ] cold and warm water

How Black Affects Radiation

Using the flame from the candle, blacken the outside of one of the cans. Be sure to wear protective gloves so that you are not burned by a hot can or dripping wax. The other can should be shiny. Put a thermometer in each can and fill each can about two-thirds full of cold water. Set the cans side by side in warm sunlight. Monitor the changes in temperature. The water in the dark can will begin to warm first. The radiant heat from the sun strikes both cans equally, but the shiny can reflects most of the heat, while the black can absorbs most of the heat.

Fig. 43-1. *Apply a dull coat of black soot to the outside of a can by holding it over a flame.*

Empty the cans and fill them about two-thirds full of warm water. Place them side by side in the shade and again watch the change in temperature. The water in the dark can will start to cool first. The warm water causes both cans to radiate heat, but heat will radiate faster from the black surface.

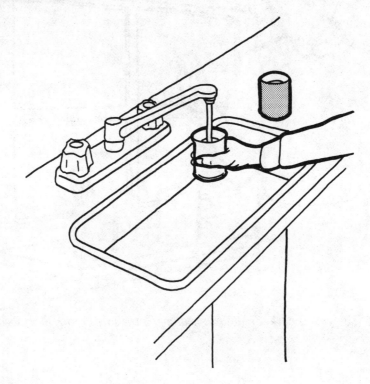

Fig. 43-2. *Fill the black can and the shiny can with water.*

Fig. 43-3. *The dark can absorbs heat while the shiny can reflects heat.*

Experiment 44

Focusing Radiant Energy

Materials

- ☐ magnifying glass
- ☐ tissue paper
- ☐ glass bowl
- ☐ sunlight

Place a small wad of tissue paper in the bottom of the bowl. The bowl is to keep the fire from spreading. Hold the magnifying glass in the sun and focus the rays to a fine point on the tissue paper. In a few seconds smoke will appear, then the paper will catch fire. The magnifying glass brought the heat rays to a point of intense heat, causing the paper to quickly reach its kindling point. Be sure not to go off and leave your project or leave the magnifying glass lying about, it could start a fire.

Fig. 44-1. *A magnifying glass can concentrate light rays and create intense heat.*

Experiment 45

Materials

- [ ] empty ink bottle with a cork stopper
- [ ] glass tube from a medicine dropper
- [ ] small tube
- [ ] large, clear glass-bowl
- [ ] very cold water
- [ ] very hot water
- [ ] ink or food coloring

Make two holes in the cork stopper the size of the two tubes (see Fig. 45-1). The medicine dropper tube should have the small end up and the other end flush with the bottom side of the cork. The small end should stick up past the cork nearly two inches. The other tube will have one end flush with the top of the cork and extend nearly to the bottom of the ink bottle.

Fill the bowl with very cold water. Color some very hot water with the ink or food coloring and pour it into the ink bottle. Press the stopper in place with the dropper tube sticking up. Wipe off any outside water and quickly place it in the bowl of cold water. A "volcano" will immediately start to erupt. The heavy, cold water will press into the tube going to the bottom of the bottle. This pushes the lighter hot water up and out the dropper tube toward the surface. This experiment visually demonstrates convection currents.

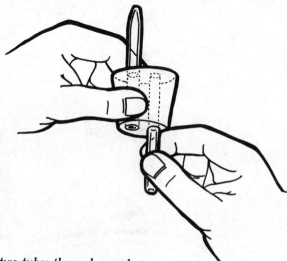

Fig. 45-1. *Insert two tubes through a cork.*

Fig. 45-2. *Fill a bowl with cold water.*

Fig. 45-3. *Pour the hot water into the ink bottle.*

Fig. 45-4. *Place the ink bottle in the bowl of cold water.*

Fig. 45-5. *An ink volcano flows from the bottle.*

Experiment 46

Heat Energy and Temperature

Materials

- ☐ nail (uncoated and non-galvanized)
- ☐ pliers
- ☐ bowl of cold water
- ☐ boiling water
- ☐ stove
- ☐ thermometer

Warning: Do not use any type of coated or galvanized nails. Heating of galvanized metal produces a poisonous gas.

Place the thermometer in the bowl of cold water and notice the temperature. Using the pliers, heat the end of the nail over the stove until it is almost glowing red. Now lower the nail into the bowl of cold water. The temperature of the water barely changed even though the nail was extremely hot. Try pouring a little boiling water into the cold water. The temperature of the cold water immediately started to rise.

The temperature of the boiling water was only 212 degrees Fahrenheit, much lower than the nail.

Fig. 46-1. *Measure the temperature of the water.*

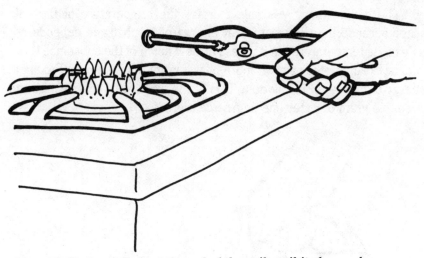

Fig. 46-2. *Carefully heat the end of the nail until it glows red.*

Fig. 46-3. *Try to heat the water with the nail. The temperature stays about the same.*

The boiling water was able to raise the temperature of the cold water because the amount of heat energy of a substance depends not only on its temperature but its mass or how much of the substance there is. The boiling water gave off more heat energy than the nail. Total heat energy is measured in calories. This is the amount of heat required to raise one gram of water one degree centigrade.

Fig. 46-4. *Add a little boiling water and the temperature instantly starts to rise.*

Experiment 47

Materials

- ☐ bowl of water
- ☐ rubbing alcohol
- ☐ electric fan

How Evaporation Lowers Temperature

Lower one hand in the bowl of water, then place it in the stream of air from a fan (or blow on it). Your hand will feel cool. Now wet your hand with the rubbing alcohol and try it again. This time your hand will feel much cooler.

This is because the heat from your hand is removed from the surface of the skin as the water or alcohol evaporates. This lowers the temperature of the skin. Water and alcohol both evaporate, so both cool, but alcohol evaporates faster so it cools more.

Fig. 47-1. *Submerge one hand in water for a few seconds.*

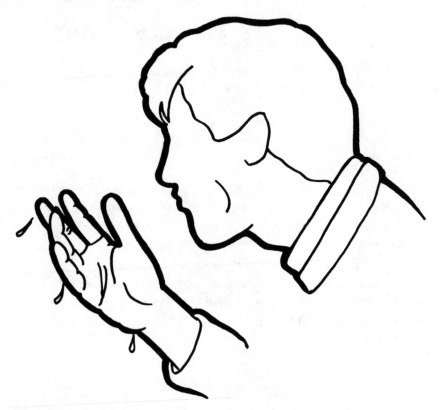

Fig. 47-2. *You hand is cooled by evaporation.*

Experiment 48

Materials

☐ candle and matches
☐ protective gloves

Heat and Convection Currents

In a warm room, open the door to the outside a few inches. Hold the candle near the top of the opening. Be sure to wear protective gloves so that hot, dripping wax won't burn your hands. The candle flame will point to the outside. This means that warm air is flowing out through the opening. Now lower the candle to the opening near the floor. The flame now points in the opposite direction showing that a cool flow of air is entering the room. As the warm air flows out the top of the open door, it allows cooler air to enter at the bottom of the door. This is a convection current and shows that the warmer, less dense air is rising, and the cooler, more dense air is falling.

Fig. 48-1. *Warm air flows out of the top of the door because warm air is less dense and rises.*

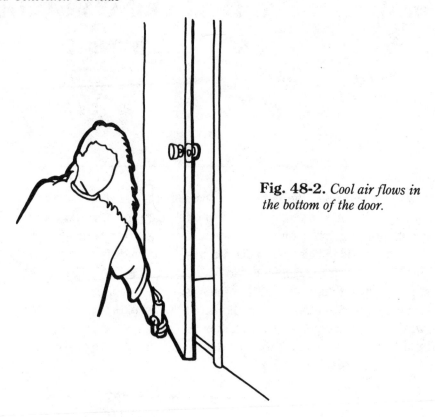

Fig. 48-2. *Cool air flows in the bottom of the door.*

Experiment 49

Chimneys and Heat

Materials

- ☐ two cardboard tubes (from roll of tissue)
- ☐ shoe box with lid
- ☐ small candle stub, mounted in a metal lid
- ☐ matches
- ☐ sheet of paper
- ☐ pencil

Using the cardboard tube as a guide, draw a circle in the lid about two inches from each end. Cut out both circles so that the two tubes will fit snugly in the holes. The ends should barely stick through the lid. These will be the chimneys. Place the candle directly under one of the chimneys. The candle must be short enough so that the flame will not burn the top of the box. Next, light the candle and place the lid on the box. The candle must be centered beneath the chimney. Twist the paper into a tight roll and light one end. After a couple of seconds blow it out. All you want is smoke. Hold the smoking paper over the chimney without the candle. Instead of rising, the smoke will go down this chimney and come out the chimney above the candle. The heat from the candle warms the air in the chimney causing it to expand. The cooler dense air is pulled down the other chimney drawing the smoke with it. This cooler air enters the box and pushes the warmer air out the chimney above the candle, taking the smoke with it.

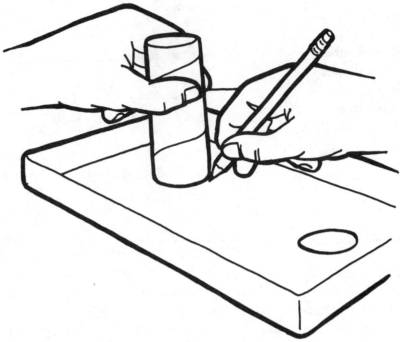

Fig. 49-1. *Mark out the openings in the lid by tracing the end of the tissue roll.*

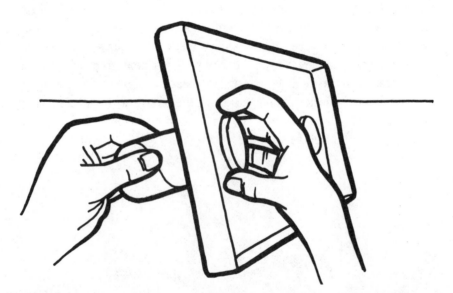

Fig. 49-2. *Fit the tubes in the openings. They should just barely fit.*

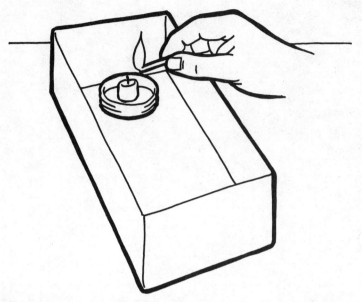

Fig. 49-3. *Place the candle in one end of the box. Align it directly beneath the hole in the lid.*

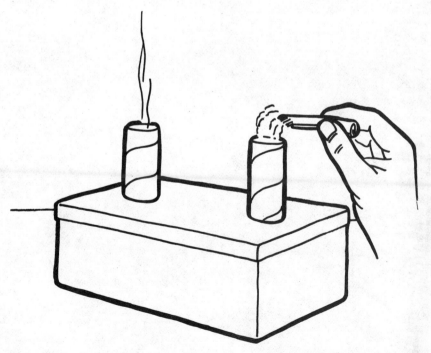

Fig. 49-4. *Smoke goes down one chimney and out of the other.*

PART II

SCIENCE FAIR PROJECTS

Science Fair Projects

A SCIENCE FAIR PROJECT CAN BE VERY EXCITING, BUT IT DOES CREATE problems. The biggest problem will probably be choosing a subject. Be sure to give this a lot of thought. If you pick a subject too quickly, you might discover you have to abandon it for another because it is too complicated, expensive, or the materials are not easily available. Put your imagination to work but keep your project within your abilities. When choosing a project, pick a topic that you're really interested in. A science fair project should make you enthused. A project that is too complicated could be beyond your ability because of the cost of materials or whether equipment is available. A simple experiment that is well demonstrated can be much more important than a complicated one that is not performed well. Often, major scientific breakthroughs are discovered using simple equipment.

Begin by dividing your science fair project into basic steps. For example:

(1) Choosing a topic

(2) Questions and hypothesis. The hypothesis is simply your guess of the results of the experiment.

(3) Doing the experiment

(4) The results

(5) Your conclusions

Normally, you will need to make a report on your experiment. It should show what you wanted to prove or a question you wanted to answer. Graphs and charts can be a helpful tool in explaining projects (see Fig. 50-1). The report should describe the experiment, the results, and the conclusions you made.

After you have selected a topic, you will want to be able to display your experiment. You might need to build a model. Often, this can be

Fig. 50-1. *Graphs and charts can help explain your project—how you collected data, your results, and conclusions.*

built from wood or cardboard. It might be made from normal throw-away items from home: empty coffee cans, plastic or glass bottles, cardboard tubes from paper towels or tissue, and empty wooden spools from sewing thread (see Fig. 50-2). Try to be creative; use your imagination.

Once you have picked the topic of your project, decide on a specific question to be answered. Don't generalize. Have a definite problem to solve or prove. For example, the topic of how temperature affects how metal expands and contracts could be narrowed down to how far a particular size steel bridge will expand or shrink between winter and summer. You could demonstrate this by constructing a scale model bridge of rigid copper tubing, then heating it with a lamp or candle and cooling it with ice cubes (see Fig. 50-3). You could even show how a bridge would buckle if it did not have expansion joints.

Fig. 50-2. *Much of the material used in a science fair project can be found around the home.*

You could use the experiment concerning the insulating abilities of wet and dry materials to compare different materials used in clothing for survival at extreme temperatures. The need to stay dry would be very important.

What happens when pressure changes the melting point of ice can be used to show why glaciers move or how this affects the way ice skate blades are designed.

You might want to use the experiment *Focusing Radiant Energy* to build a solar oven. This can be made by assembling small panels covered with smooth aluminum foil into a parabolic (bowl-like) mirror. A black tin can could be placed in the center of the mirror as a receiving container. The material to be heated is placed in the black can. The parabolic mirror focuses the sun's ray on the can.

Fig. 50-3. *A copper tube can be used to show how bridges shrink and expand.*

You could set this up on a table in front of a cardboard panel. The ends of the panel could be folded forward so that it stands by itself, similar to the back part of a theater stage. If the project is displayed indoors, you could suspend a light as a substitute sun and two thermometers could show the temperatures inside and outside of the oven (see Fig. 50-4).

The panels displaying the experiment can show the information compiled in your report, the purpose of your experiment, construction of various types of solar ovens, and the results and conclusions of your experiment. This could include the efficiency of solar cooking and some of the places it might be used. For example, where other fuels are scarce.

Fig. 50-4. *Demonstrating a solar oven.*

Any of the experiments in this book can be done in more detail and used as a science fair project. With little imagination, a simple experiment can be expanded and developed to a point where it will be very interesting and enlightening. Most any experiment has been done before, but maybe you can approach it from a different point of view. Heat has been around since the beginning of time, but there are still important discoveries to be made.

Index